THE EYE BLACK STUDY

Dr. Brian M. DeBroff

DG
DUNLAP GODDARD

Copyright © 2013 by Dr. Brian M. DeBroff, M.D., F.A.C.S.

All rights reserved. No part of this book may be reproduced, stored in a retrieval system, scanned, or distributed in any printed or electronic form by any means without permission. Please do not participate in or encourage piracy of copyrighted materials in violation of the author's rights. Purchase only authorized editions.

Published in the United States of America
Manufactured in the United States of America

ISBN-13: 978-0-965-6040-5-5 (trade paperback: acid free)
ISBN-13: 978-0-965-6040-6-2 (ebook)
LCCN: 2013934286

Dr. Patricia Pahk's Yale doctoral thesis is reprinted with the author's permission.

Article from the *Archives of Ophthalmology* reprinted with the author and journal's permission.

Jacket Design by Jeremy Bronson
Cover Photo: Tim Shaw of the Tennessee Titans by Ronnie Kadykowski

QUANTITY PURCHASES
Companies, professional groups, clubs, and other organizations may qualify for special terms when ordering quantities of this title. For information, email Special Sales Department at info@dunlap-goddard.com.

A portion of the proceeds from each book sold will go to Distressed Children & Infants International's Childhood Blindness Prevention Program and to the University of Michigan's C.S. Mott Children's Hospital, one of the nation's leading pediatric institutions known around the world for its excellence.

Contents

Forewords — 5

Introduction — 7

Author's Note — 9

Archives of Ophthalmology: The Ability of Periorbitally Applied Antiglare Products to Improve Contrast Sensitivity in Conditions of Sunlight Exposure — 15

Doctoral Thesis: Ability of Periorbitally Applied Antiglare Products to Improve Contrast Sensitivity in Conditions of Sunlight Exposure — 27

Forewords

It gives me great pleasure to introduce this innovative book that illustrates the dramatic impact academic research can have on our daily lives. In this study, antireflective black grease was shown to improve contrast sensitivity and reduce glare in athletes subjected to bright sunlight exposure. With the reprinting of Dr. Patricia Pahk's thesis (and Dr. Brian DeBroff's detailed author's note), the reader has a clear understanding of how this research study was conceptualized and undertaken. Moreover, the collaboration between Dr. Pahk and Dr. DeBroff illustrates Yale School of Medicine's emphasis on self-directed learning and exploration and originality in its student research.

I wish to congratulate Dr. DeBroff and Dr. Pahk on the publishing of this book.

James C. Tsai, M.D., M.B.A.
Robert R. Young Professor and Chairman
Department of Ophthalmology and Visual Science
Yale University School of Medicine
Chief of Ophthalmology, Yale–New Haven Hospital

We would like to thank Dr. DeBroff for his friendship over the years and for the diligent work he put into this book. Without the expertise, time, and wisdom of both he and Patty Pahk, our family company would not exist today. We pride ourselves on making the best eye black for athletes.

We know the science in this book will benefit athletes on the field. As a legacy to our late father and grandfather, Andy Farkas, we hope and pray that this book and our company will do great things off the field as well. We are proud to support the great work done by Dr. Brian M. DeBroff and his nonprofit organization, Distressed Children & Infants International (DCI).

Dr. DeBroff and Executive Director Dr. Ehsan Hoque founded DCI in 2003. As a company committed to helping athletes see better on the field, we can think of no greater calling than to support DCI's Blindness Prevention Program, helping the world see better off the field.

For more information on this organization please visit:
http://www.distressedchildren.org

Thank you,
The Farkas Family

Introduction

The *Archives of Ophthalmology* published "The Ability of Periorbitally Applied Antiglare Products to Improve Contrast Sensitivity in Conditions of Sunlight Exposure"—also known as "the Eye Black Study"—in 2003. This is the first time this article and the study on which it is based have been published together in their entirety.

The publication of the Eye Black Study generated a significant amount of attention. Athletes in many sports employ various products to combat glare and this study dared to ask a simple question: Does any of this stuff really work? The results, as it turns out, surprised many of us.

As I sit here, recounting the twists and turns of the last ten years I can hardly believe that the study included here has led not only to the founding of Farkas Eye Black, but also my placement on board of Farkas Eye Black, with the likes of Brian Rooney, whose family founded the Pittsburgh Steelers, investors like Gary Gigot, and the grandchildren of Andy Farkas.

Every game day in America and around the world, thousands of professional and amateur players, coaches, parents, and equipment managers now turn to Farkas Eye Black. It is a fitting legacy for Andy Farkas—the man who changed the face of the American athlete.

I am very proud to be a part of this team, and very proud that science led to the creation of a family company making their product proudly in the USA.

Dr. Brian M. DeBroff M.D., F.A.C.S.
Yale Medical School
New Haven, Conn.

8 / The Eye Black Study

Author's Note

In 1999, a bright and eager medical student, Patricia Pahk, approached me with interest in a research topic in ophthalmology for her medical school thesis. After I suggested a number of topics, she chose a project to determine the effectiveness of antiglare products used by amateur and professional athletes. This topic is important because so many athletes now wear grease or stickers—athletes who are role models for today's youth. And among those who wear stickers, the fashion trend is toward those that feature catchphrases, advertisements, or even biblical references. In addition, given Yale University's unique connection to the start of football, I thought it was the perfect opportunity to answer a question relevant to both ophthalmology and sports medicine.

I came to this investigation with the belief that a player could not reduce glare and improve contrast sensitivity simply by smearing a substance or applying stickers on their cheekbones. I saw eye black and stickers as nothing more than psychological war paint—a fad with no scientific basis. To test this hypothesis, Patty and I designed an experiment to determine the performance of the various products employed by athletes against a control while answering the question of whether or not antiglare products offer any visual benefit to athletes.

The process started with gaining the approval of Yale's Human Research Protection Program for the use of human subjects. We decided to use a Pelli-Robson Contrast Sensitivity Chart. Unlike a standard eye chart, the letters on the Pelli-Robson chart remain the same size on each line but grow fainter from top to bottom. The experiment we designed would test if either of the antiglare products improved the study group's ability to correctly identify progressively fainter lines under conditions of sunlight exposure.

On the day of the testing, the conditions were ideal—a clear, summer day with not a cloud in the sky. The testing was performed at a park on the medical school campus, which was unobstructed by trees. Forty-six students volunteered to participate in the experiment. Patty had gathered the students from Yale's schools of Nursing, Epidemiology, Public Health, and Medicine.

The group was quite impressive; 91 percent of them had actively participated in sports. Thirteen percent even played at the varsity level in college and a few reported at least one past use of eye black.

The study group consisted of twenty-four women and twenty-two men between the ages of eighteen and thirty. No one wearing glasses was selected for the study due to the effect of lenses on the perception of glare.

Together, we conducted the experiment while volunteers assisting us in collecting the demographic data on each participant. We randomized the participants to application and testing of Vaseline grease, eye black grease, and eye black stickers. Each eye was tested separately on the charts. The team recorded nearly three hundred readings.

Patty compiled the results in her thesis: *Ability of Periorbitally Applied Anti-glare Products to Improve Contrast Sensitivity in Conditions of Sunlight Exposure*. The results of our investigation were thorough and solid—the findings statistically and clinically significant:

- Eye black stickers offer no improvement over the control.
- Eye black grease significantly improved contrast sensitivity over control.

The completed doctoral thesis was about fifty pages. It described the purpose and methods of her experiment and its unexpected results. I was not surprised when Patty successfully defended her thesis; her conclusions were grounded on solid research and rigorous statistical analysis.

Patty also included in her thesis an exhaustive and extensive history of glare reduction in sports, as well as the naturally occurring facial marks of various animals. She also discovered that in 1942, Washington Redskins fullback Andy Farkas was the first professional football player to use eye black.

When it came to writing an article for publication in the ophthalmic literature, Dr. Pahk suggested that I write the article due to her time constraints as an ophthalmology resident and my extensive experience with writing scientific articles. Together we prepared a scientific paper that was peer reviewed and published in the prestigious *Archives of Ophthalmology*. The public relations office at Yale University recognized that the sports aspect of this study would be of interest to a wider audience and announced the publication of "the Eye Black Study."

I have more than fifty published scientific articles to my credit—many describing pioneering techniques of pediatric cataract and intraocular lens techniques—but none ever produced the kind of media frenzy generated by the Eye Black Study. Within the week of publication, ESPN, CNN, NPR, the *Los Angeles Times*, *Science* magazine, and many others called to interview me.

During an interview with the *New York Times*, I learned that the owner of a company that makes eye black stickers was questioning the motivation, methods, and results of our study. The reporter informed me that this owner was very upset that the study found sticker products ineffective at reducing glare; he had derided the athleticism of our test subjects and suggested that replicating on-field conditions might yield a different result.

The owner of the sticker company quickly commissioned a rival study to prove the effectiveness of stickers in reducing glare—and he got what he paid for. In this study, laser beams were shot at mannequin heads and those with stickers detected reduced glare. It is unclear if the mannequin head study employed a rigorous scientific methodology or even how mannequin heads in a lab replicated the conditions an athlete might experience on the field.

The results of the Eye Black study have been replicated and confirmed by researchers at the University of New Hampshire. Although I was skeptical at first, I have become a believer in eye black grease and would certainly wear it myself before taking the field. I feel that I have made a contribution to sports to the benefit of youngsters who look to professional athletes as role models.

The most significant phone call I received came from Brian Farkas, the grandson of NFL eye black pioneer Andy Farkas. Brian and his sister, Katie Gates, wanted to know all the details of the tradition and the science behind what their late grandfather had created, and a friendship quickly developed. The brother-sister team decided to start a company based on the science and invited me to serve as their scientific adviser.

From the very beginning, they indicated that anything with their grandfather's name on it had to be the best, and not just the product, but the company behind the product. They were in the early stages of recruiting a board of directors that would supplement their talents.

Soon, Brian invited me to join a board of directors for his company that included Brian Rooney—a board member of the Pittsburgh Steelers—and Gary Gigot—a Seattle-based investor who funded the Center for Entrepreneurial Studies at the University of Notre Dame.

With funding in place, I began working with chemists and scientists from Ann Arbor, Michigan—based Avomeen Analytical Services to formulate the best eye black in the world. We went through several test batches, and a few were close, but the Farkas family demanded that only the best would get the green light for the initial production run. They would not rush to market in time for the approaching football season. They felt they had only one shot at this so they had to get the product right.

The right formula emerged from the lab and survived intense testing. The first 2,500 tins of Farkas Eye Black quickly sold around the country and soon the company heard back from its customers. Duke, Virginia, and top college and high school teams noticed a difference and demand surged.

The Ability of Periorbitally Applied Antiglare Products to Improve Contrast Sensitivity in Conditions of Sunlight Exposure

Brian M. DeBroff, M.D.; Patricia J. Pahk, M.D.
Published July 2003 *Archives of Ophthalmology*

Background: Sun glare decreases athletes' contrast sensitivity and impairs their ability to distinguish objects from background. Many commercial products claim to reduce glare but have not been proven effective in clinical studies.

Objective: To determine whether glare-reducing products such as eye black grease and antiglare stickers reduce glare and improve contrast sensitivity during sunlight exposure.

Design and Methods: We tested 46 subjects for contrast sensitivity using a Pelli-Robson contrast chart. Each subject served as an internal control and then was randomized to either application of eye black grease, antiglare stickers, or petroleum jelly at the infraorbital rim. All testing was performed in conditions of unobstructed sunlight.

Results: Analysis of variance revealed a significant difference between eye black grease (mean ± SD, Pelli-Robson value, 1.87 ± 0.09 log MAR units) and antiglare stickers (1.75 ± 0.14 log MAR units) in binocular testing ($P = .02$). No statistical difference was found between the groups in right eyes, left eyes, or in combined data from the right and left eyes. Paired t tests demonstrated a significant difference between control (mean ± SD, 1.77 ± 0.14 log MAR units) and eye black grease (1.87 ± 0.09 log MAR units) in binocular testing ($P = .04$). There was also a significant difference between control (mean ± SD, 1.65 ± 0.05 log MAR units) and eye black grease (1.67 ± 0.06 log MAR units) in combined data from the right and left eyes ($P = .02$).

Conclusion: Eye black grease reduces glare and improves contrast sensitivity in conditions of sunlight exposure compared with the control and antiglare stickers in binocular testing.

LIGHT DAMAGES eye structures as a result of the physical phenomenon of energy transmission. Light also has a psychophysical component that affects the quality of vision. Scattering of light can produce glare, which in turn can lead to visual disability. Athletes are particularly challenged by the effects of light radiation and glare. Glare from sunlight or stadium lighting impairs an athlete's contrast sensitivity and impairs the ability to see detail if the light source is from elsewhere in the visual field.[1] Scattered light degrades contrast sensitivity by splashing extra, non-information-containing light onto the retinal image and reducing the contrast of the image.[2] Studies have shown that the higher an athlete's contrast sensitivity, the more likely the athlete can discriminate an object as its velocity increases.[3]

Natural protection from glare is provided by facial anatomy, including the brow and forehead, the bony orbital cavity, cheekbones, and the upper and lower eyelids. The ocular media has protective light absorbing and reflecting properties. In addition, now there are many commercial products that balance UV protection with glare reduction; many of these items, however, have not been proven effective in clinical studies and may provide a false sense of protection.

The first known glare-reducing devices were made by Eskimos from Alaska, Canada, and Siberia approximately 2,000 years ago. Ivory or wooden goggles with horizontal slots effectively allowed peripheral vision while blocking out light reflected by snow and ice. The Chinese used colored transparent pebbles gathered from riverbeds for protection. The earliest recorded use of "sports sunglasses" is attributed to Nero, who viewed gladiators through an emerald. More recently, Tuberville, a 15th-century English ophthalmologist, prescribed silk veils for his post-operative patients complaining of photophobia, and, in 1886, the mail order company Sears, Roebuck, and Company began to offer sunglasses.[4-6]

Currently, available glare reducers include visors, sunglasses, and contact lenses. In addition there are various filters for glare reduction, including photochromic lenses, polarizing filters, and tinted filters. Also, antireflective coating can be placed on lenses to reduce glare. Sunglasses, however, can lower background illumination and diminish visual acuity, especially at dimmer levels of light.

Eye black, a form of face paint applied to the cheekbone, is a controversial product that has been used by athletes for decades to reduce sun glare. It is thought to reduce reflected glare into athletes' eyes from the cheekbone by absorbing incident light with its dark pigment. Eye black grease is made from a mixture of beeswax, paraffin, and carbon. Antiglare stickers are made from patented fabric. According to product advertisements, the correct positioning is one-half inch below the eyelid on the cheekbone, and the reported function is similar to that of the natural masks found on wolves, badgers, and even killer whales.[7-8]

Professional baseball and football players have been using eye black for decades, and other sports are beginning to catch on. More recently, antiglare stickers have become available commercially. The history of eye black is unknown; there is no history of the product anywhere in the annals of baseball, and its obscure arrival has become part of the folklore of the game. The first photographic evidence of its use is found in a 1942 photograph of Washington Redskins' football player Andy Farkas in a game against the Philadelphia Eagles. At the time, evidence suggests that players used to burn cork and then smear the ashes on their cheeks.[7]

The actual effectiveness of eye black has been a constant source of debate, in part because no trials have ever been performed in its decades-long history. Curt Mueller, owner and president of Mueller Sports Medicine (Prairie du Sac, Wisconsin), has been selling it for nearly 40 years but has never seen any studies proving its effectiveness.[7] Eye black has become a sports accessory, with players donning it during night games and indoor games. Athletes use it for the competitive edge, an aggressive look, and an extra psychological advantage.

The purpose of this study is to determine if glare-reducing products such as eye black grease and antiglare stickers marketed to athletes for reduction of glare actually improve contrast sensitivity during sunlight exposure. To do this, we designed a randomized, controlled trial using natural sunlight as our source of glare. We used the Pelli-Robson contrast sensitivity chart to document changes in contrast sensitivity before and after randomization to 1 of 3 treatment groups: eye black grease, antiglare stickers, and petroleum jelly placebo.

Methods

We recruited 46 students (92 eyes) for a 1-time measurement of contrast sensitivity using a Pelli-Robson contrast chart. Each student served as an internal control by initially being tested without a product and then tested again after being randomized to application of eye black grease (Eye Blackgrease; Easton Sports, Inc., Van Nuys, California), petroleum jelly, or an antiglare (No Glare; Mueller Sports Medicine, Inc.) sticker. Each product was applied by the same data collector on the participant's skin at the level of the infraorbital rim just prior to testing. A second Pelli-Robson contrast chart with a different order of optotypes was used to avoid familiarity upon retesting.

Participants were students drawn from the schools of Medicine, Nursing, and Epidemiology and Public Health of Yale University, New Haven, Connecticut. The Yale University School of Medicine institutional review board approved the project and informed consent forms, and informed consent was obtained for all participants. Students wearing eyeglasses were not included in the study because of the effect of the lenses on glare and contrast sensitivity.

Testing was conducted outdoors during a period of direct and unobstructed sunlight, with the nearest trees more than 50 yards away. All subjects faced into the sun during a 3-hour period from noon to 3 p.m. A volunteer assigned each subject a number and collected demographic data, including age, sex, ethnicity, a brief ocular history, the level of sports participation, and previous use of eye black or antiglare stickers. Contrast sensitivity testing was performed by placing the subject approximately 1 m away from a Pelli-Robson chart. Subjects were asked to read as far as they could using each eye separately and then with both eyes together. No time limits were set, and subjects were asked to guess letters they felt they could not see. The Pelli-Robson contrast sensitivity value was determined by the last set of triplets with 2 or more letters correctly read and was recorded in log MAR (logarithm of the minimal angle of resolution) units. Data consistency was maintained by having a separate data collector for each Pelli-Robson chart who remained for the duration of the study.

Statistical analysis was performed using ANOVA (analysis of variance) and paired *t* tests. The ANOVA testing was done to compare results between the treatment groups, whereas paired t tests tested for a difference between the control and each treatment group. Raw data included data from the right eye, the left eye, both eyes, and the combined data from the right and left eyes. $P<.05$ was considered statistically significant. Data are given as mean ± SD.

Results

Subjects were aged from 18 to 30 years (mean, 23 years). Women composed 52.2% of the subject population; 62.8% were white, 32.6% were Asian, and 4.7% were Hispanic. Contact lenses were worn in the past by 54.4%, and 8.9% reported some type of ocular history, including corneal abrasions and dry eyes. A history of sports participation was elicited from 91% of subjects, and 8.7% reported having used eye black or antiglare stickers at least once.

From a total of 46 participants, 276 readings were recorded (testing each eye separately and then both eyes together for each of the 2 Pelli-Robson charts). Of the 46 students, each was tested as a control; 16 were assigned to the eye black grease treatment group, 16 to the petroleum jelly treatment group, and 14 to the antiglare sticker group.

An ANOVA was used to compare differences between the treatment groups of eye black grease, petroleum jelly, and antiglare stickers (Table 1). There was a significant difference between eye black grease (Pelli-Robson value, 1.87 ± 0.09 log MAR units), petroleum jelly (1.78± 0.11 log MAR units), and antiglare stickers (1.75 ± 0.14 log MAR units) in students tested binocularly ($P = .02$). A Bonferroni multiple comparison test was performed to test pairwise difference within this group and demonstrated that the statistically significant difference was between eye black grease and antiglare stickers. There was no significant difference found between the 3 groups in the right or left eye alone or in the combined data from right and left eyes ($P = .86$, $P = .59$, and $P = .55$, respectively).

Table 1. Pelli-Robson Values for Analysis of Variance

	Mean ± SD, LogMAR Units			
Data Source	Eye Black Grease (n = 16)	Petroleum Jelly (n = 16)	Antiglare Sticker (n = 14)	P Value
Right eye	1.67 ± 0.05	1.66 ± 0.07	1.65 ± 0.14	.86
Left eye	1.67 ± 0.08	1.64 ± 0.09	1.64 ± 0.11	.59
Binocular	1.87 ± 0.09	1.78 ± 0.11	1.75 ± 0.14	.02
Combined right and left eye*	1.67 ± 0.06	1.64 ± 0.08	1.64 ± 0.13	.55

Abbreviation: logMAR, logarithm of the minimum angle of resolution.
*Sample sizes for combined data are eye black grease, n = 32; petroleum jelly, n = 32; antiglare sticker, n = 28.

A paired t test was used to compare the control with each treatment group (Table 2). We demonstrated a statistically significant difference between the control (Pelli-Robson value, 1.77 ± 0.14 log MAR units) and the eye black (1.87± 0.09 log MAR units) in binocular testing (P =.04). We also found a statistically significant difference between the control (1.65 ± 0.05 log MAR units) and the eye black group (1.67 ± 0.06log MAR units) by combining the data from the right and left eyes (P = .02). There was no statistically significant difference between control and eye black grease in the right or left eye alone (P = .16 and P = .08, respectively). There was no statistically significant difference between control and petroleum jelly in the right eye, left eye, binocularly, or with combined data from the right and left eyes (P = .50, P = .58, P = .27, and P = .37, respectively). There was no statistically significant difference between control and antiglare stickers in the right eye, left eye, binocularly, or with combined data from the right and left eyes (P =.19, P = .58, P = .58, and P = .16, respectively).

Table 2. Control Pelli-Robson Values for Paired *t* Tests

Data Source	Eye Black Grease Group (n = 16)	P Value†
Right eye	1.65 ± 0.00	.16
Left eye	1.64 ± 0.07	.08
Binocular	1.77 ± 0.14	.04
Combined right and left eye*	1.65 ± 0.05	.02

Abbreviation: logMAR, logarithm of the minimum angle of resolution.
*Sample sizes for combined data are eye black grease group, n = 32; petrc
†P values are for control vs treatment. Treatment values are presented in 1

Comment

In the between-groups analysis, we found a statistical difference between eye black grease and antiglare stickers in binocular testing. Although there were statistically significant differences in the contrast sensitivity, the actual differences on the Pelli-Robson chart testing varied. The actual mean Pelli-Robson contrast sensitivity value was 1.87 ± 0.09 log MAR units for eye black and 1.75 ± 0.14 log MAR units for antiglare stickers. This is about equivalent to 1 level of contrast sensitivity difference on the Pelli-Robson chart, which decreases in contrast in equal logarithmic steps of 0.15. Similarly, in analysis between control and treatment, we found a statistically significant difference in the eye black group in both binocular testing and the combined data from the right and left eye. For binocular testing, this translated into a control mean Pelli-Robson value of 1.77 ± 0.14 log MAR units and an eye black mean value of 1.87 ± 0.09 log MAR units. Again, this is about equivalent to 1 level of contrast sensitivity difference. However, the combined data from the right and left eyes had a control mean value of 1.65 ± 0.05 log MAR units and an eye black mean value of 1.67± 0.06 log MAR units, a much smaller difference in contrast sensitivity. The reason behind the values was apparent on the testing day. We observed a significant decrease in recognition at contrast levels of 1.65 log MAR units, particularly in monocular testing. We urged participants to continue guessing until we had recorded a value but certainly observed that subjects consistently had difficulty at a similar point on the chart. As a result, monocular testing revealed mean values clustered around 1.65 with very small SDs regardless of treatment group. Binocular testing fared better with most participants reading beyond the 1.65 level.

What is the significance of 1 level of improved contrast sensitivity? In a study of the test-retest reliability of the Pelli-Robson chart, Elliot et al.[9] measured contrast sensitivity in the dominant eye of normal younger and older populations. Most of the younger population (mean age, 22.5 years) were found to have a Pelli-Robson mean value of 1.80 log units or better. Retesting 2 weeks later showed contrast sensitivity scores to be repeatable to within 0.15 log units (or the equivalent of 1 step in contrast sensitivity). Thus, they define a significant change as a difference of 0.30 log units, or 2 steps on the Pelli-Robson chart. According to this study, 1 level of improved contrast sensitivity in the same subject—that is, in those who tested in the control group and the eye black treatment group—is within the error margin of the test reliability and not necessarily a result of glare-reducing products. The study by Elliot et al. differed from our own in 3 major respects: all testing was done monocularly and without a glare source and tests occurred 2 weeks apart. Possibly as a result of these differences, we did not see the same pattern in our data. Each participant in our study underwent a test-retest situation by testing as a control and in a treatment group, with a unique Pelli-Robson chart at each reading. There was a significantly smaller difference between the first and second readings in those who were randomized to the petroleum jelly and antiglare sticker groups, both of which had minimal antiglare effects. As a result, our data seem to show higher test-retest reliability, and the differences found between the control and eye black groups in the binocular testing becomes more significant in our study. However, both the study by Elliot et al. and our study agree that testing for smaller gradations in contrast sensitivity would improve the results.

The Pelli-Robson contrast sensitivity chart was chosen because of its high test reliability and ease of function. There are 2 main approaches to testing contrast sensitivity. The first is a subjective method that uses pattern testing with printed or electronically generated charts. The second is the objective or electrophysiologic method, which measures pattern visual evoked potentials. However, this method is too time consuming for clinical purposes. Test-retest reliability is the highest with familiar optotypes such as the letters in the Pelli-Robson chart or Landolt C rings.[10] Letters have the advantage of having innate orientation and minimizing the odds of guessing correctly. Letter charts are also relatively error tolerant because the subject is typically allowed to make 1 mistake per line without affecting the results. As a result, there is a growing interest in using more familiar optotypes to measure contrast sensitivity. First-time subjects tend to be conservative in their answers; requiring them to guess improves measuring accuracy.[11] We standardized our testing protocol by having a single tester test all subjects for each chart. Each tester asked the subject to continue guessing until the test recorded 2 incorrect responses within a triplet.

Prior to testing, preliminary sample size calculations, based on an α of .5 and power of 80, indicated that we would need a sample size of 40 subjects in each treatment group to have significant results. Our sample size was much smaller, with a total of 46 participants. As a result, it was unclear whether our results would achieve the normal distribution necessary for parametric analysis. We analyzed the data with both parametric and nonparametric tests: parametric analysis was done with ANOVA and paired t tests; equivalent nonparametric tests were the Kruskal-Wallis 1-way ANOVA and the Wilcoxon matched-pairs signed-rank test. Our results were nearly identical in either case, indicating that our small sample size nevertheless reflected a normal distribution. Thus, we present the data here with parametric analysis.

It is impossible to avoid a certain level of bias associated with the application of the various products. Eye black is prevalent enough in our culture that participants could guess the intended effects. There was no way to mask the participants to their treatment groups. We also suspect a learning bias based on repeated readings of the chart. Subjects were asked to read both charts with each eye separately and then binocularly. In all, each subject gave 6 readings, 3 for each chart. We attempted to control for this possibility by providing charts with different letters.

However, we found that binocular scores were consistently higher than monocular scores; whether this is the effect of longer exposure to the chart, which resulted in clarification of hard-to-see letters, or the superiority of binocular contrast sensitivity is difficult to assess.

Our study had a slight majority of women, which introduces the question as to whether it would be possible that eye black benefits male athletes. Brabyn and McGuinness[12] have shown there are sex differences in contrast sensitivity, with women being more sensitive to low spatial frequencies and men more sensitive to high spatial frequencies but no significant difference being seen in midrange spatial frequencies. Our study itself was too small to give any reliable information on sex breakdown.

There is questionable validity in combining right and left eye data to double the sample size. It has been shown that left and right eye readings are not independent readings, which stems from the fact that they arise from the same single brain. In terms of contrast sensitivity testing, binocular contrast sensitivity has been shown to have higher sensitivity than monocular sensitivity across all spatial frequencies compared. The difference was shown to be approximately 42% higher than the predicted sum of the monocular responses.[13-14] Gilchrist and McIver also showed that decreased luminance in one eye from any ocular condition decreases the binocular contrast sensitivity such that it is worse than the better eye.

Other limitations of the study include the inability to measure sunlight luminance and the position of the sun. Thus, it would be difficult to relate our results with future results achieved under similar conditions. Variability in ambient testing conditions could affect our results by exposing participants to varying levels of sun brightness. Perceptual brightness changes with the angle of the sun, with cloud conditions, and with time of year. We performed our testing within a 3-hour period on a single day to minimize variability in sun luminance; however, gradual changes could not be assessed.

In summary, we found eye black grease to be statistically superior to control and to antiglare stickers in 3 situations. There was a statistically significant difference between eye black grease and antiglare stickers in binocular testing. There was also a statistically significant difference between the control and eye black grease in binocular testing and in the combined data from the right and left eyes.

Based on this study, eye black grease appears to be more than psychological war paint. These results suggest that eye black grease does in fact have antiglare properties, whereas antiglare stickers and petroleum jelly do not. Perhaps the mixture of wax and carbon in eye black grease is superior for reducing reflected light than is the fabric material in antiglare stickers.

The cheekbone itself reduces glare by reflecting light away from the eye socket. Placing a pigment on top of the cheekbone could theoretically absorb more light. Future studies may help elucidate the best location and material for maximal improvement of contrast sensitivity. The greatest challenge facing contrast sensitivity measurement and glare testing is the lack of standardization in procedures, both in stimulus parameters and testing style. Future studies may benefit from a control-led glare source, larger sample sizes, and more sensitive contrast sensitivity testing techniques.

Corresponding author and reprints: Brian M. DeBroff, M.D., Department of Ophthalmology and Visual Science, Yale University, 330 Cedar St, P.O. Box 208061, New Haven, CT 06520-8061 (e-mail: brian.debroff@yale.edu).

References

1. Rubin GS, Clinical glare testing. In: Schachat AP, ed., *Current Practice in Ophthalmology*. St Louis: Mosby–Year Book Inc., 1992; 153–63.
2. Miller D, Sanghvi S, Contrast sensitivity and glare testing in corneal disease. In:Nadler MP, Miller D, Nadler DJ, eds., *Glare and Contrast Sensitivity for Clinicians*. New York: Springer-Verlag, 1990; 45–52.
3. Kluka DA, Love PL, Kuhlman J, Hammach G, Wesson M. The effect of a visual skills training program on selected collegiate volleyball athletes. *Int J Sports Vision*. 1996; 323–34.
4. Charman WN, MacEwen CJ, Light and lighting. In: Loran DFC, MacEwen CJ., eds., *Sports Vision*. Oxford: Butterworth-Heinemann, 1995; 88–112.
5. Miller D. The effect of sunglasses on the visual mechanism. *Surv Ophthalmol*. 1974; 1938–44.

6. Miller D, Nadler MP, Light scattering: Its relationship to glare and contrast in patients and normal subjects. In: Nadler MP, Miller D, Nadler DJ, eds., *Glare and Contrast Sensitivity for Clinicians*. New York: Springer-Verlag, 1990; 24–32.
7. Bryce R. Out of left field. *Austin Chronicle*, 2001. Available at: http://www.austinchronicle.com/issues/vol18/issue44/xtra.leftfield.html . Accessed February 5, 2002
8. Glareblox stick-on strips product website. Available at: http://www.glareblox.com/intro.html. Accessed February 5, 2002.
9. Elliot DB, Sanderson K, Conkey A. The reliability of the Pelli-Robson contrast sensitivity chart. *Ophthalmic Physiol Opt*. 1990; 1021–24.
10. Lempert P, Standards for contrast acuity/sensitivity and glare testing. In: Nadler MP, Miller D, Nadler DJ, eds., *Glare and Contrast Sensitivity for Clinicians*. New York: Springer-Verlag, 1990; 5–23.
11. Wolfe JM, An introduction to contrast sensitivity testing. In: Nadler MP, Miller D, Nadler DJ, eds., *Glare and Contrast Sensitivity for Clinicians*. New York: Springer-Verlag, 1990; 113–19.
12. Brabyn LB, McGuinness D., Gender differences in response to spatial frequency and stimulus orientation. *Percept Psychophys*. 1979; 26319–324.
13. Leege GE, Binocular contrast summation, I: detection and discrimination. *Vision Res*. 1984; 24373–383.
14. Gilchrist J, McIver C. Fechner's paradox in binocular contrast sensitivity. *Vision Res*. 1985; 25609–613.

Author Information
Submitted for publication November 14, 2002; final revision received February 23, 2003; accepted March 11, 2003.

This study was presented at the Annual Meeting of the Association for Research in Vision and Ophthalmology, May 8, 2002, Fort Lauderdale, Florida.

Drs. De Broff and Pahk had full access to all the data in the study and take responsibility for the integrity of the data and the accuracy of the data analysis.

ABILITY OF PERIORBITALLY APPLIED ANTI-GLARE
PRODUCTS TO IMPROVE CONTRAST SENSITIVITY IN
CONDITIONS OF SUNLIGHT EXPOSURE

A Thesis Submitted to the
Yale University School of Medicine
In Partial Fulfillment of the Requirements for the
Degree of Doctor of Medicine

by Patricia J. Pahk

2002

ABILITY OF PERIORBITALLY APPLIED ANTI-GLARE PRODUCTS TO IMPROVE CONTRAST SENSITIVITY IN CONDITIONS OF SUNLIGHT EXPOSURE.

Patricia J. Pahk and Brian M DeBroff. Department of Ophthalmology and Visual Science, Yale University School of Medicine, New Haven, CT

Sun glare decreases an athlete's contrast sensitivity and impairs his ability to distinguish objects from background. Many commercial products claim to reduce glare but have not been proven effective in clinical studies. We determined whether glare-reducing products such as EyeBlack grease and No Glare sticker reduce glare and improve contrast sensitivity during sunlight exposure.

46 students were tested for contrast sensitivity using a Pelli-Robson Contrast Chart. Each subject served as an internal control and was initially tested using no product. Each was then tested again after being randomized to either application of EyeBlack grease (n=16), No Glare sticker (n=14), or Vaseline (placebo; n=16) on the skin at the level of the infraorbital rim. All testing was performed in conditions of direct and unobstructed sunlight exposure to the subject. ANOVA test revealed a significant difference between EyeBlack grease (Pelli-Robson value 1.87 +/- 0.09) and No Glare sticker (1.75 +/- 0.14) in binocular testing ($p = 0.0182$). No statistical difference was found between the groups in the right eye, left eye, or in the combined data of the right and left eyes. Paired T-test demonstrated a significant difference between control (1.77 +/- 0.14) and EyeBlack grease (1.87 +/- 0.09) in binocular testing ($p = 0.0364$). There was also a significant difference between control (1.65 +/- 0.05) and EyeBlack grease (1.67 +/- 0.06) in the combined data of the right and left eyes ($p = 0.0208$). There was no statistical difference found between control and treatment group in any other combination.

Our data suggests that EyeBlack grease reduces glare and improves contrast sensitivity in conditions of sunlight exposure as compared to control and to No Glare sticker in binocular testing. Further testing with larger sample sizes and controlled glare conditions are needed to determine if glare reduction occurs in a manner that would reduce sun glare in actual athletic conditions.

TABLE OF CONTENTS

Abstract	28
Figures and Tables	30
Acknowledgements	31
Introduction	32
Methods	44
Results	46
Discussion	51
Conclusion	60
References	63

FIGURES

Figure 1	Glare degradation	34
Figure 2	Contrast sensitivity function *(CSF)* curve	36
Figure 3	Example of Pelli-Robson contrast chart	47
Figure 4	Two sets of Pelli-Robson chart optotypes used in this study with corresponding logMAR values	48
Figure 5	Point vs. diffuse glare source	55

TABLES

Table 1	Mean Pelli-Robson values with standard deviation for ANOVA—original data	49
Table 2	Mean Pelli-Robson values with standard deviation for ANOVA—combined right and left eye data	49
Table 3	ANOVA p-values	49
Table 4	Mean Pelli-Robson values with standard deviation for Paired T-test original data	50
Table 5	Mean Pelli-Robson values with standard deviation for Paired T-test combined right and left eye data	50
Table 6	Paired T-test p-values	50

APPENDICES

Appendix 1	EyeBlack survey questionnaire	61
Appendix 2	Pelli-Robson contrast chart data collection sheet	62

ACKNOWLEDGEMENTS

I would like to thank Dr. Brian DeBroff for his guidance, his enthusiasm, and his patience over the past two years.

I would also like to extend my gratitude to the following people: To the Offices of Student Research and Student Affairs. To Joan Buonconsejo for her time and assistance with the statistical analysis. To the many individuals in the medical, nursing, and EPR schools, some who helped me organize the study and others who participated in it. Finally, I would like to thank my family for their constant support.

INTRODUCTION

Light damages eye structures as a result of the physical phenomenon of energy transmission. Light also has a psychophysical component that affects the quality of vision. Scattering of light can produce glare, which, in turn, can lead to visual disability. Athletes are particularly challenged by the effects of light radiation and glare. There are many commercial products which balance ultraviolet (UV) protection with glare reduction; many of these items, however, have not been proven effective in clinical studies and may provide a false sense of protection to the eye. EyeBlack, a form of facepaint applied on the cheekbone, is one such controversial product that has been used by athletes for decades to reduce sun glare.

Athletic performance is clearly influenced by the athlete's visual function. Studies to date indicate that different visual skills are important for performance in different sports and that athletes and nonathletes differ in these visual skills. Stereopsis, visual reaction time, dynamic visual acuity, and contrast sensitivity, to name a few, are superior in athletes as compared to nonathletes. Professional athletes perform better in these visual functions than nonprofessional athletes, who perform better than normal populations (1–4). Furthermore, Solomon et al. (5) showed that stereoacuity is better in major league batters than pitchers, indicating that visual skills relate to athletic requirement. Kioumourtzoglou et al. (6) tested members of the Greek national teams of basketball, volleyball, and water polo and also determined that the demands of each sport select for specific visual abilities.

Contrast sensitivity is useful to all players because it helps athletes distinguish objects from their backgrounds under varying lighting conditions. Laby et al. (3) tested binocular contrast sensitivity in 387 major and minor league professional baseball players using the Vistech 6500 Vision Contrast Test System, Contrast Sensitivity Viewer–IOOO contrast sensitivity testing system, and Binocular Visual acuity tester. A comparison of baseball players' scores with those of a normal population showed that athletes had greater contrast sensitivity, particularly at high spatial frequencies. In addition, their study indicated that major league players had better contrast sensitivity than minor league players, though the sample sizes were sometimes too small to reach statistical significance. Laby points out that minor league players represent a combination of players on their way to the majors as well as those who remain in the minors and would require an even larger sample size to reach significance.

Melcher and Lund (7) tested dominant eye contrast sensitivity as well as lowlight- and glare-affected vision in 232 high school student athletes. They compared the eight-member girls' volleyball team that reached state tournament level to the general high school athlete (n=224), the female athlete sample (n=78) and other female volleyball players (n=46). Contrast sensitivity was measured using the Vistech wall chart at testing light level of 30-40 ft/l. Low light vision was tested using the Night Sight Meter, which measures the relative ability of an individual to identify the orientation of a *20/300* Landolt C target in a testing apparatus as surround lighting is dimmed. Glare testing was done using the same apparatus but with the addition of dazzle provided by a light adjacent to fixation. The girls' volleyball team tested significantly better in these visual skills compared to the general athlete. They tested relatively better, though not always significantly so, compared to other volleyball players and female athletes.

Similarly, Hoffman et al. (8) used Arden grating plates to show an increased ability to detect small differences in contrast in collegiate baseball players compared to controls. However, few studies have demonstrated how increased contrast sensitivity or changes in light levels may affect athletic performance. In a study of volleyball players, Kluka et al. (9) found that the higher the athlete's contrast sensitivity, the more likely the athlete can discriminate an object as its velocity increases. Campbell et al. (10–12) found that batter reaction times to pitches are slower in low light levels, a particular handicap against major league 90 mph pitchers. Relatively simple solutions exist for such conditions: postpone play or turn on stadium floodlights. A more difficult condition arises when daytime or stadium light levels are very high. Glare from sunlight or stadium lighting impairs an athlete's contrast sensitivity. Major league fielders are selected for their ability to keep track of objects against the sun but, even so, most use a variety of glare-reducing products to improve their contrast sensitivity and performance in conditions of disabling glare. We found no studies that show how glare may affect contrast sensitivity in athletes.

Glare

The Illuminating Engineering Society defines glare as the discomfort or impairment of vision experienced when parts of the visual field are excessively bright in relation to the general surroundings. Glare can cause visual discomfort (*discomfort glare*), for example, when sun is reflected off sand or snow. It may impair ability to see detail (*disability glare*) if a light source is elsewhere in the visual field, such as in the case of oncoming headlights (13).

Disability glare is caused by any factor that would scatter light in the ocular media. Even normal young eyes scatter 10–20% of incident light. Of this, it is estimated that 70% is scattered at the lens and approximately 30% at the cornea. Normal vitreous scatters less than 1% of incident light, though vitreous opacities will significantly increase the amount of light scattered (13). The term "glare degradation" (Figure 1) refers to the fact that scattered light degrades contrast sensitivity by splashing extra, non-information-containing light onto the retinal image and reducing the contrast of the image (14). It has been shown that eliminating wavelengths less than 450nm improves image quality by reducing chromatic aberration and light scattering without affecting color (15).

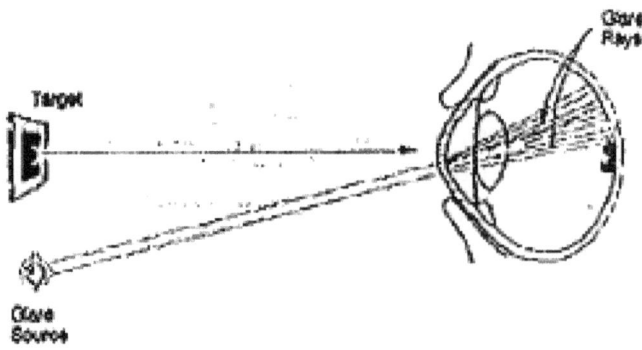

Figure 1: *glare degradation* (14)

There are many factors that influence how much solar radiation and potential glare reaches the eye. Before UV radiation reaches the eye, it is largely absorbed by ozone, carbon dioxide, and water vapor in the earth's atmosphere. Changes in the ozone are concerning because the ozone typically filters out UV-C light and absorbs some UVA and UV-B rays. Increasing the distance that light has to travel increases the likelihood that shorter wavelengths (UV, blue light) will be scattered before it reaches the earth's surface. Thus, seasonal studies have shown a 17-fold increase in solar radiation intensity in summer than in winter. Day studies show a 10-fold increase in at noon than three hours before or after. Likewise, altitude can be a factor; elevations of 1,000 meters increases radiation intensity 16% (16). Local weather factors can also affect the amount of radiation reaching the ground. Clouds do not preferentially filter out short wavelength light because they are made up of water droplets, which scatter all wavelengths equally (17). Finally, it should be noted that most people do not look directly at the sun because of glare discomfort. However, glare can also result from rays reflected from below. Fresh snow reflects about 80% of incident radiation, older snow over 50%, clean white sand 30%, water 5%, earth and grass less than 5% (18).

Contrast Sensitivity

Contrast sensitivity is defined as the ability to detect the presence of minimal luminance differences between objects and is the reciprocal of the contrast threshold (19). Contrast is generally calculated as a percent:

$$\frac{\text{Max intensity} - \text{min intensity}}{\text{Max intensity} + \text{min intensity}} \times 100$$

Visual acuity is the upper limit of spatial frequency below which an item cannot be resolved regardless of its contrast; in essence, it tests only one dimension. Snellen charts are fixed at near 100% contrast. However, all objects have size and contrast and thus occupy a two-dimensional space. The relationship between contrast and size is known as the contrast sensitivity function (CSF). Normal CSF shows greatest sensitivity to test patterns of intermediate size; sensitivity decreases as the patterns grow smaller or enlarge. On this curve, visual acuity is the point where the CSF crosses the x-axis (Figure 2) (20). This leads to the natural question whether visual acuity can predict contrast sensitivity. Melcher's study (7) noted a strong relationship between visual acuity and contrast sensitivity, particularly in the high frequencies. Bodis-Wollner and Diamond have shown that decreasing size and contrast result in the same high frequency cut-offs (equivalent to visual acuity) but different CSF curves. Thus, Snellen visual acuity cannot predict the shape of CSF.

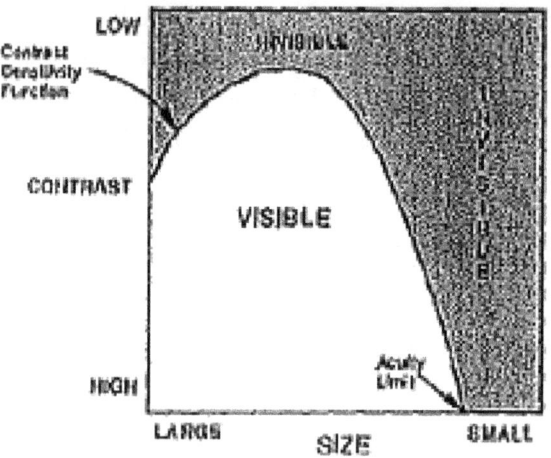

Figure 2: contrast sensitivity function (CSF) curve (20)

The shape of a CSF curve is a product of optical, retinal, and neural factors. Optical and photoreceptor characteristics provide the primary limitations for high frequency reductions in sensitivity. The eye has a high frequency limitation because of inherent optical aberration. The density of photoreceptors roughly matches the optical limits of the eye. Low frequency drop off is due to spatial antagonism within the retina. Retinal ganglion cells receive responses from an area of photoreceptors (known as the receptive field of the cell). Ganglion cell receptive fields have a characteristic "center-surround" organization, where stimulation of the photoreceptors in the center causes an increase in the cell's response but stimulation of the sUl1-ound is inhibitory. Stimulation by a large patch of light will activate both center and surround and will produce relatively little activation of the ganglion cell. Cell response will be greatest when a bright bar falls on the center of the receptive field while a flanking dark bar falls on the inhibitory regions. Thus there will be an optimal spatial frequency in the medium range with cell response falling off as the frequency becomes higher or lower. This output then proceeds to the lateral geniculate nucleus of the thalamus and from there to the visual cortex. Contrast sensitivity function is roughly a compilation of different ganglion cells with varying receptive field sizes (20, 23).

There are many different variables that affect contrast sensitivity. Contrast sensitivity is proportional to mean luminance and the size of the stimulus. Retinal location is important since contrast sensitivity declines as the image moves away from the fovea. Refractive states and pupil size affect the quality of the retinal image and therefore contrast sensitivity. Pupil size has two primary effects: retinal light level varies with pupil size and large pupils introduce larger optical aberrations due to corneal irregularities. Intermediate pupil sizes (2–5mm) provide the best acuity and contrast sensitivity. Contrast sensitivity has a temporal component; for example, at low frequency, it is likely to be enhanced by abrupt stimulus onset (20, 24). Contrast sensitivity also varies with the orientation of stimuli and Appelle (25) showed that Caucasians have decreased sensitivity to oblique orientations compared to vertical or horizontal stimuli (known as the oblique effect).

OVERVIEW OF EYE PROTECTION IN SPORTS

Human protection against the disabling effects of sun glare has both evolutionary components and learned behavior. Learned behaviors include blinking, squinting, and turning our heads. Natural protection is provided by facial anatomy, including the brow and forehead, the bony orbital cavity, cheekbones, and the upper and lower eyelids. The ocular media has protective light absorbing and reflecting properties. Control is found even at the level of the muscle through pupillary miosis. Pupil constriction varies according to the intensity of illumination. Experiments show that the pupil does most of its size changing under dimmer levels of illumination, ranging from full dilation to *slightly* less than 3mm as *light* illumination changes from 10^{-6} lumens (dark night) to 10 lumens. From 10 lumens to 10,000 lumens (bright day), the pupil only decreases to about 2 mm, a comparatively smaller change (16, 26).

The first known glare-reducing devices were made by Eskimo from Alaska, Canada, and Siberia as long as 2,000 years ago. Ivory or wooden goggles with horizontal slots effectively allowed peripheral vision while blocking out light reflected by snow and ice. The Chinese might have used colored transparent pebbles gathered from riverbeds for both magnification and light protection. The earliest recorded use of "sports sunglasses" is attributed to Nero, who viewed gladiators through an emerald, presumably to absorb *light* radiation in the amphitheater. More recently, Turberville, a fifteenth-century English ophthalmologist, prescribed silk veils for his postoperative patients complaining of photophobia, and, in 1886, the mail order catalogue company Sears, Roebuck, and Company began to offer sunglasses (16, 26, 27).

Currently, available glare reducers include visors, sunglasses, and contacts. A rapidly growing field is in the development of various filters for glare reduction. Filters can be integrated into spectacles, goggles, and visors and vary from tint to polarizing filters. EyeBlack, a controversial product, is hypothesized to absorb light with its dark pigment and reduce sun glare reflecting into the eyes.

Visors

Visors and large-brim hats are used not only to reduce glare but also to make surfaces look brighter. This is illustrated by the analogy of a tunnel. Generalized bright light makes surfaces appear relatively dark (as a tunnel appears on approach); wearing a visor effectively places the wearer inside the tunnel. However, visors do not protect against light reflected from below and are estimated to reduce ocular exposure by half.

Sunglasses

Sunglasses are widely preferred for athletes because of the major advantage of protection from injury. The effectiveness of sunglasses depends on the level of sun intensity. In fact, one doesn't see better with sunglasses, only more comfortably. Brightness discrimination and visual acuity definitely improve as illumination increases; in dim light, sunglasses would lower background illumination and diminish visual acuity. At high levels of illumination, much more information is coming in from which one must discriminate the actual image from the "noise." The pupil is more effective than sunglasses in a dimly lit room but not in a bright outdoor environment (26).

Filters

Absorptive lenses control the concentration and the distribution of the incident radiation. When light passes through a lens, some is reflected at the front, some is absorbed or scattered by the lens material, and some is reflected at the back surface of the lens. The mechanism by which tinted lenses improve contrast is essentially because the image must travel the lens three times (across each layer of tint, lens, and tint), thus greatly reducing their luminance. The percentage of light that passes through the lens at a given wavelength is the *lens transmittance* for that wavelength. The average of transmittance values over a wavelength range is the *mean transmittance*. Thus, filters can be described in terms of their mean transmittance over the visible, UV, and infrared (IR) spectrum.

Photochromic Lenses

Photochromic lenses are active filters, with transmission characteristics that vary with ambient light. This is achieved by means of silver halide crystals embedded in the lens. High levels of short-wavelength light cause the molecules to partially dissociate and give off free silver, so that the lenses become dark and their transmittance is reduced. The process reverses when light levels fall, and the lenses clear. Commonly the basic tint is neutral gray or brown and the transmittance varies from about 20% to 80% (16). These lenses are often used in combination with antireflective coating to reduce glare.

Polarizing Filters

Light reflected from smooth horizontal surfaces, such as water surface or the hood of a car, tend to be predominantly horizontally polarized. A linear polarizing filter which transmits only vertically polarized light will suppress these reflections. Since only one plane of polarization is transmitted by a linear polarizer, such filters can transmit a maximum of 50% of normal unpolarized light, and usually only transmit approximately 40% (16). In addition to reducing glare, they effectively act as neutral filters. Water sports enthusiasts and fishers are able to see objects below the surface of the water more easily. Bikers, runners, and drivers have the combined advantage of reducing glare off the road, car hood, or wet surfaces.

Tinted Filters

The best all-purpose tint is a neutral gray because it will not distort color perception while absorbing 98% of UV and IR waves. Green absorbs 99% of UV and 100% of IR waves but tends to shrink the width of the spectrum; it allows in green light, which the human eye is most sensitive to, and absorbs red and blue light. It is also less effective in blocking bright light. Brown lenses have similar properties to green but has even greater color distortion because the lens absorbs more blue light (31).

Yellow or amber lenses have 100% UV absorption but transmit IR and 83% of the visible spectrum (31); these transmit only longer-wavelength light and exclude the blue end of the spectrum. The shorter blue wavelengths are likely to be scattered more in the ocular media than longer wavelengths, so filtering blue wavelengths will improve retinal image contrast. Fishermen often use yellow-tinted lenses in combination with polarizers because the lens absorbs the blue light scattered off the surface of water and the fish underneath become more clearly visible (16). However, evidence does not support the argument that yellow filters improve visual acuity or reduce glare. Some studies have shown that yellow filters may enhance performance on tasks involving the detection of low-contrast targets at intermediate spatial frequencies (16). This could justify the use of such filters by hunters, skiers, and boaters in low-visibility conditions, such as scattered light in fog and haze.

In contrast, yellow filters have been shown to improve the ability to detect subtle variation in the contour of the snow for skiing. Skiers use a narrow range of lens colors (yellow, orange, amber, rose) to make out shadows and bumps in the snow and to know where to turn. The bottoms of holes (illuminated from the zenith of the sky) may appear bluer than the tops (illuminated by the horizon sky) so that a yellow filter will enhance the contrast of the hole. Light transmitted through the snow, which is preferentially absorbed at longer wavelengths, may also contribute to the apparent blueness of the bottom of snow holes (16). Skiers also use polarized lenses which filter out glare and reflected light, eliminating "bounce back" of sun off snow and ice.

Recently, blue tinted lenses have been marketed as enhancers of visual performance in tennis. The blue tint will enhance contrast between a yellow tennis ball and the background by filtering through the natural fluorescence of the yellow tennis ball, which is blue green rather than yellow. By reflecting yellow light, the actual yellow of the tennis ball is filtered out. However, Marmor (32) points out that highlighting the fluorescence of the ball removes the contrast of the yellow ball and the green surface court (grass, hard, and clay) and instead places a blue-green ball against the green court. Only against red clay courts would the contrast of the blue-green tennis ball be advantageous. Further, blue tinted lenses are poor UV absorbers and allow near UV light in. Blue light is less effective with respect to functions such as acuity, contrast detection, and motion perception; critical perception is actually reduced. Blue light may also contribute to macular aging and degeneration. In addition, by blocking out the brighter 13 colors, the pupil is usually relatively dilated, allowing even more exposure to short wave light (32). As a result, athletes are discouraged from using blue tinted lenses.

Safety Standards

There is no governing body that dictates quality control of eye products. Instead, a variety of guidelines are published by interested groups of experts that assess both the technical as well as the consumer preferences (28, 29). A sample set of guidelines is included below: (adapted from Zagelbaum (29))

1. UV B (280–320) less than 5% transmittance, less than 1% transmittance for less than 310nm.

2. UV A (320–400) less than 10% transmittance and absolutely less than maximal visible light transmittance.
3. Blue light (400–500) less than 10% transmittance and absolutely less than the maximal visible light transmittance. 25–50% transmittance desirable.
4. Long wavelength visible light (500–760) less than 15% transmittance for bright conditions, such as sand or snow.
5. Infrared (above 760) filtration desirable but not essential.
6. Side shields (e.g., wraparound frames) and either a rim across the top or used in conjunction with a brimmed hat to protect against oblique incident radiation in very bright conditions.
7. Optional polarization to prevent glare.

Coatings

There are two common forms of coating that are placed on lenses. The first, antireflective coating, is designed to reduce lens glare by means of an ultrathin layer of magnesium fluoride. The maximum transmittance of a clear lens is limited by the reflectance at the lens surface; as a result, the upper limit of transmittance of any lens could not be greater than 92%. The antireflective coating decreases reflectance and increases transmittance to about 99% (30). The second coating is a minor coating that is best suited for situations of intense glare (snow, water). Glass lenses with 20% transmittance coated with a mirror coating of 40% transmittance will have a combined transmittance of 8% (30).

Contact Lenses

Contact lenses are not as popular in sports settings because wind, dry air, and decreased oxygen at high altitudes make them difficult to wear in many activities. More importantly, they fail to provide the protection that sunglasses offer in high impact sports. Contact lenses are available in tints (different from colored aesthetic lenses that change the color of the iris) and with UV coating. However, they can only provide UV protection over the surface they cover. Contact lens wearers are advised to wear sunglasses to protect the rest of the eye.

EyeBlack

EyeBlack, and products like it, are thought to reduce reflected glare into athletes' eyes from the cheekbone by absorbing incident light with its dark pigment. EyeBlack grease is made from a mixture of beeswax, paraffin, and carbon; stickers are made from patented fabric. According to product advertisements, the correct positioning is one-half inch below the eyelid on the cheekbone. The same advertisement reports that the function of EyeBlack is similar to that of the natural masks found in nature as seen on wolves, badgers, and even killer whales (33, 34).

Professional baseball and football players have been using EyeBlack for decades and other sports are beginning to catch on. The history of EyeBlack is unknown; there is no history of the product anywhere in the annals of baseball and its obscure arrival has become part of the folklore of the game. The first photographic evidence of its use is found in a 1942 photograph of Washington Redskins football player Andy Farkas in a game against the Philadelphia Eagles. At the time, evidence suggests that players used to burn cork and then smear the ashes on their cheeks (33).

The actual effectiveness of EyeBlack has been a constant source of debate in part because no actual trials have ever been done in its decades-long history. Mueller of Mueller Sports Medicine has been selling it for nearly 40 years but has never seen any studies proving its effectiveness (33). EyeBlack has become a sports accessory, with players donning it during night games and indoor games. Athletes use it for the competitive edge, an aggressive look, and an extra psychological advantage.

The purpose of our study is to determine if glare-reducing products such as EyeBlack grease and No Glare stickers marketed to athletes for reduction of glare actually improves contrast sensitivity during sunlight exposure. To do this, we designed a randomized, controlled trial using natural sunlight as our source of glare. We used the Pelli-Robson contrast sensitivity chart to document changes in contrast sensitivity before and after randomization to one of three treatment groups: EyeBlack grease, No Glare stickers, and Vaseline placebo.

METHODS
Overview

46 students for a total of 92 eyes were recruited for a one-time measurement of contrast sensitivity using a Pelli-Robson Contrast chart (Figure 3). Student participants were tested first without product and then tested again after being randomized to application of Easton brand EyeBlack grease, Vaseline grease, or Mueller brand No Glare sticker. EyeBlack grease and No Glare sticker are two examples of products currently marketed to reduce sun glare and enhance athletic performance.

Study Participants, Eligibility, and Exclusion Criteria

Participants were students drawn from the schools of Medicine, Nursing and Epidemiology and Public Health of Yale University during an orientation activity held at a park in August 2001. Students were asked to participate in the study during breaks in their activity. Students wearing eyeglasses were not included in the study because of the effect of the lenses on the perception of glare.

Testing Conditions

Testing was conducted in the middle of a grass park outlined by a lightly pigmented concrete walkway on one side and a line of trees on the other. We tested all participants between noon and 3 p.m. on a single day in August 2001. The subjects faced east, into the sun with their backs to the trees, which were approximately ten yards behind them. The concrete walkway lie about five yards beyond the Pelli-Robson chart. The nearest trees the subjects faced were approximately 50 yards away. Sunlight luminance, either overhead or at eye level, and the position of the sun in the horizon were not recorded.

Data Collection

A volunteer assigned each student a number and collected demographic data including age, gender, ethnicity, brief ocular history, level of sports participation, and previous use of EyeBlack. All testing was performed outdoors in conditions of direct and unobstructed sunlight exposure to the subject. Testing was completed within a three-hour period from noon to 3 p.m. Visual acuity was not assessed due to the time limitations of the study. Students wearing glasses were excluded from the study due to glare effects of the lenses but otherwise allowed with corrective contact lenses. Pupil size was not measurable with a pupillometer due to the bright conditions.

Each student served as internal control and was initially tested using no product. The subject was then tested again after being randomized to either application of EyeBlack grease, Vaseline grease, or No Glare sticker. Each product was applied by the same data collector on the participants' skin at the level of the infraorbital rim just prior to testing. A second Pelli-Robson Contrast chart with a different order of optotypes was used to avoid familiarity upon retesting.

Testing was done by placing the student approximately 1 meter away from a Pelli-Robson chart. The student was asked to read as far as he could using each eye separately and then with both eyes together. No time limits were placed on the student and students were encouraged to "guess" letters they felt they could not see. The Pelli-Robson contrast sensitivity value was determined by the last set of triplets with 2 or more letters correctly read and was recorded in logMAR (minimal angle of resolution) units (Figure 4). Data consistency was maintained by having a separate data collector for each Pelli-Robson chart who remained for the duration of the study.

Data Entry

Demographic data was distributed to study participants in survey form (Appendix 1) and entered into Excel. Participants' contrast sensitivity was recorded by each data collector (Appendix 2) and entered into Excel.

Statistical Analysis

Statistical analysis was performed using ANOVA and paired T test. ANOVA testing was done to compare results between the treatment groups while paired T tests tested for a difference between control and each treatment group. Raw data included data from the right eye, the left eye, both eyes, and the combined data of the right and left eyes. A p-value of less than 0.05 was determined to signify a statistically significant result (rejection of the null hypothesis).

RESULTS

Demographics

Student demographics were as follows:
- age ranged from 18 to 30 years with a mean of 23 years
- 52.2% were female
- 62.8% Caucasian, 32.6% Asian, and 4.7% Hispanic
- 54.4% wear contacts
- 8.9% report some ocular history including corneal abrasions and dry eyes
- 91% report some level of sport participation
- High school: Junior varsity (JV) 17.4%, Varsity (V) 50%
- College: JV 6.5%, V 13%, intramural 56.5%
- Recreational: 60.9%
- 8.7% report having used EyeBlack at least once in the past

Data Analysis

276 readings were recorded from a total of 46 student participants (testing each eye separately and then both eyes together for each of the two Pelli-Robson charts). Of the 46 students, each tested as a control, 16 were assigned to the EyeBlack grease treatment group, 16 to the Vaseline treatment group, and 14 to the No Glare sticker group.

We used ANOVA to compare differences between the treatment groups of EyeBlack grease, Vaseline grease, and No Glare sticker (Tables 1, 2, 3). We determined that there is significant difference between EyeBlack grease (mean Pelli-Robson value 1.87 +/- 0.09), Vaseline grease (1.78 +/- 0.11), and No Glare sticker (1.75 +/- 0.14) in students tested binocularly ($p = 0.0182$). A Bonferroni multiple comparison test was performed to test pairwise difference within this group and demonstrated that the statistically significant difference was between EyeBlack grease and sticker. There was no significant difference found between the three groups in the right or left eye alone or in the combined data of the right and left eyes ($p = 0.8607$, $p = 0.5932$, and $p = 0.5509$, respectively).

We then used paired T-test to compare control to each treatment group (Tables 4, 5, 6). We demonstrated a statistically significant difference between the control (1.77 +/- 0.14) and the EyeBlack (1.87 +/-

0.09) in binocular testing (p = 0.0364). We also found a statistically significant different between the control (1.65 +/- 0.05) and the EyeBlack (1.67 +/- 0.06) by combining the data from the right and left eyes (p = 0.0208). There was no statistical difference between control and EyeBlack grease in the right or left eye alone (p = 0.1639 and p = 0.0825, respectively). There was no statistical difference between control and Vaseline in the right eye, left eye, binocularly, or with the combined data of the right and left eyes (p = 0.4973, p = 0.5805, p = 0.2702, and p = 0.3741, respectively). There was no statistical difference between control and No Glare stickers in the right eye, left eye, binocularly, or with the combined data of the right and left eyes (p = 0.1894, p = 0.5830, p = 0.5830, and p = 0.1610, respectively).

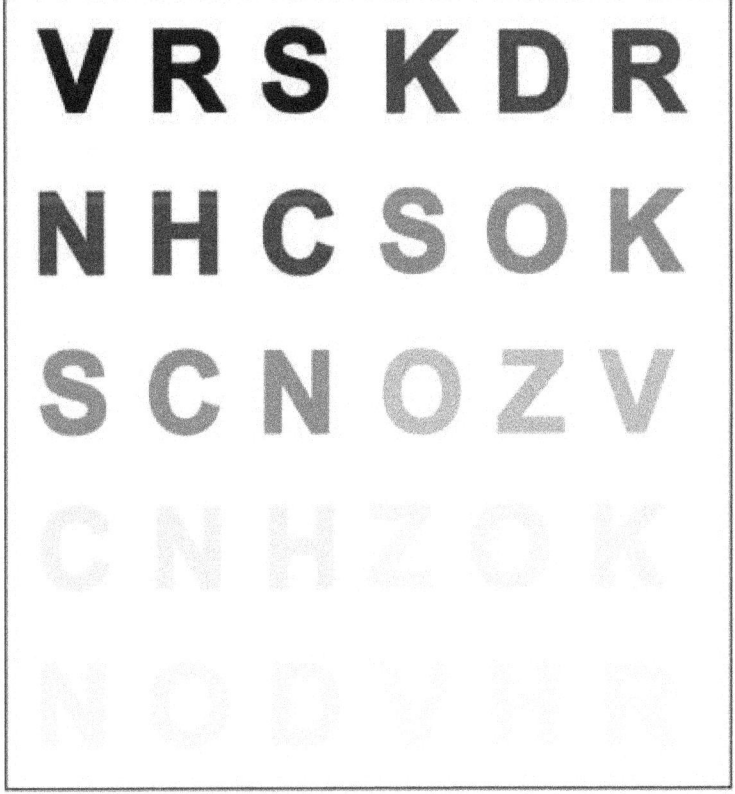

Figure 3: example of a Pelli-Robson contrast chart

0.00			0.15
0.30			0.45
0.60			0.75
0.90			1.05
1.20			1.35
1.50			1.65
1.80			1.95
2.10			2.25

0.00			0.15
0.30			0.45
0.60			0.75
0.90			1.05
1.20			1.35
1.50			1.65
1.80			1.95
2.10			2.25

Figure 4: two sets of Pelli-Robson chart optotypes used in this study with corresponding logMAR values

TABLE 1
Mean Pelli-Robson values with standard deviation for ANOVA—original data (in logMAR units)

	Right Eye	Left Eye	Binocular
EyeBlack grease (N=16)	1.67 (0.05)	1.67 (0.08)	1.87 (0.09)
Vaseline grease (N=16)	1.66 (0.07)	1.64 (0.09)	1.78 (0.11)
No Glare sticker (N=14)	1.65 (0.14)	1.64 (0.11)	1.75 (0.14)

TABLE 2
Mean Pelli-Robson values with standard deviation for ANOVA—combined right and left eye data (in logMAR units)

	Combined right and left eye data
EyeBlack grease (N=32)	1.67 (0.06)
Vaseline grease (N=32)	1.64 (0.08)
No Glare sticker (N=28)	1.64 (0.13)

TABLE 3
ANOVA p-values

	P value
Right eye	0.8607
Left eye	0.5932
Binocular	0.0182
Combined right and left eye data	0.5509

TABLE 4
Mean Pelli-Robson values with standard deviation for paired T-test—original data (in logMAR units)

	Right Eye	Left Eye	Binocular
Control—EyeBlack grease (N=16)	1.67 (0.05)	1.67 (0.08)	1.87 (0.09)
Control—Vaseline grease (N=16)	1.66 (0.07)	1.64 (0.09)	1.78 (0.11)
Control—No Glare sticker (N=14)	1.65 (0.14)	1.64 (0.11)	1.75 (0.14)

TABLE 5
Mean Pelli-Robson values with standard deviation for paired T-test—combined right and left eye data (in logMAR units)

	Combined right and left eye data
Control—EyeBlack grease (N=32)	1.67 (0.06)
Control—Vaseline grease (N=32)	1.64 (0.08)
Control—No Glare sticker (N=28)	1.64 (0.13)

TABLE 6
Paired test p-values

	Control vs. EyeBlack	Control vs. Vaseline	Control vs. sticker
Right eye	0.1639	0.4973	0.1894
Left eye	0.0825	0.5805	0.5830
Binocular	0.0364	0.2702	0.5830
Combined right and left eye data	0.0208	0.3741	0.1610

DISCUSSION

We have designed a study to test whether EyeBlack grease and No Glare stickers reduce glare and improve contrast sensitivity during sunlight exposure. In a randomized, controlled trial, we measured participants' contrast sensitivity using the Pelli-Robson chart before and after application of one of three treatment groups: EyeBlack grease, No Glare sticker, and Vaseline placebo. We have found EyeBlack to be statistically superior to control and to No Glare sticker in three situations. There is a statistically significant difference between the EyeBlack grease and No Glare sticker in binocular testing. There is also a statistically significant difference between the control and EyeBlack grease in binocular testing and in the combined data of the right and left eyes.

While there was statistical significant difference in the contrast sensitivity, the actual difference on the Pelli-Robson chart varied. In the between-groups analysis, we found a statistical difference between EyeBlack grease and No Glare sticker in binocular testing. The actual mean Pelli-Robson contrast sensitivity value was 1.87 +/- 0.09 for EyeBlack and 1.75 +/- 0.14 for No Glare sticker. This is about equivalent to one level of contrast sensitivity difference on the Pelli-Robson chart, which decreases in contrast in equal logarithmic steps of 0.15. Similarly, in analysis between control and treatment, we found a statistical difference in the EyeBlack group in both binocular testing and the combined data points of the right and left eye. For binocular testing, this translated into a control mean Pelli-Robson value of 1.77 +/- 0.14 and an EyeBlack mean value of 1.87 +/- 0.09. Again, this is about equivalent to one level of contrast sensitivity difference. However, the combined data of the right and left eyes had a control mean value of 1.65 +/- 0.05 and EyeBlack mean value of 1.67 +/- 0.06, a much smaller difference in contrast sensitivity. The reason behind the values was apparent on the testing day. We observed a significant drop off in recognition at contrast levels of 1.65, particularly in monocular testing. We urged participants to continue guessing until we had recorded a value, but certainly felt that participants consistently had difficulty at the same point on the chart. As a result, monocular testing show mean values clustered around 1.65 with very small standard deviations regardless of treatment group. Binocular testing fared better with most participants reading beyond the 1.65 level.

What is the significance of one level of improved contrast sensitivity? In a study of the test-retest reliability of the Pelli-Robson chart, Elliot et al. (35) measured contrast sensitivity in the dominant eye of normal younger and older populations. The majority of the younger population, mean age of 22.5 years, were found to have a Pelli-Robson mean value of 1.80 log units or better. Retesting two weeks later showed contrast sensitivity scores to be repeatable to within 0.15 log units (or the equivalent of one step in contrast sensitivity). Thus, they define a significant change as a difference of 0.30 log units, or two steps on the Pelli-Robson chart. According to this study, one level of improved contrast sensitivity in the same subject—that is, in those who tested in the control and EyeBlack treatment group—is within the error margin of the test reliability and not necessarily a result of glare-reducing products. The Elliot study differed in three major respects from our own: all testing was done monocularly, without a glare source, and two weeks apart. Possibly as a result of these differences, we did not see the same pattern in our data. Each participant in our study underwent a test-retest situation by testing as a control and in a treatment group, with a unique Pelli-Robson chart at each reading. There was a significantly smaller difference between the first and second readings in those who were randomized to the Vaseline and No Glare sticker groups, both of which had minimal anti-glare effects. As a result, our data seems to show higher test-retest reliability and the differences found between the control and EyeBlack group in the binocular testing becomes more significant in our study. However, both their study and ours feel that testing for smaller gradations in contrast sensitivity would improve the results.

EyeBlack, somewhat to our surprise, does appear to have antireflective properties. The cheekbone itself reduces glare by reflecting light away from the eye socket. Placing a pigment on top of the cheekbone could theoretically absorb more light. It would be interesting to see whether certain facial structures benefit more from EyeBlack or if contrast would be enhanced if the entire socket was pigmented to absorb light coming in from all directions.

It is difficult to explain why No Glare stickers performed so poorly compared to placebo and to EyeBlack. These results may very well be an artifact of sample size. It may also be because Vaseline is a poor placebo and actually has some antireflective properties. In our study, neither Vaseline nor No Glare stickers were particularly effective antiglare devices.

Choosing a Measurement of Contrast Sensitivity

We chose to use the Pelli-Robson Contrast Sensitivity chart because of its familiarity to participants, its high test-retest reliability, and its ease of function. There are two main approaches to testing CSF. The first is the subjective, or psychophysical, method, which uses pattern testing with printed or electronically generated charts. The second is the objective, or electrophysiologic method, which measures pattern visual evoked potentials (PVEPs). However, this method is too time consuming for clinical purposes. Most work in the last 20 years has involved sinusoidal, or bar, gratings. Sine waves have properties that make them attractive stimuli for measuring contrast sensitivity. First, Fourier's theorem holds that any black-and-white shape can be constructed from sine waves of appropriate frequency, contrast, phase, and orientation. The visual system essentially adds up sine waves to create that image. Sine wave gratings are simpler than letters because they represent only fundamental frequencies, whereas letters are complicated waveforms. Another desirable trait is that defocusing reduces the contrast of a sine wave but does not alter its form, as opposed to a blurred letter on a Snellen chart, which changes its appearance as well as its contrast (20, 23). However, Snellen charts have been around for 125 years and to most people, reading letters has a familiar association with eye exams. Test-retest reliability has been shown to be highest with familiar optotypes such as the letters in the Pelli-Robson chart or Landolt C rings (36). Letters have the advantage of having innate orientation and minimizing the odds of guessing correctly. Letter charts are also relatively error-tolerant since the subject is typically allowed to make one mistake per line without affecting the results (36). As a result, there is growing interest in using more familiar optotypes to measure contrast sensitivity.

Among the earliest letter charts were modified Snellen charts, each in a different contrast. Using multiple charts, both size and contrast were plotted to give an estimate of the CSF. However, there are not enough charts to test for smaller gradations in contrast sensitivity at very high and very low spatial frequencies. A variation was also done on Landolt C rings with different background contrasts and similar advantages and disadvantages. Here, the subject was asked to identify the orientation of the C. Pelli and Robson proposed a chart designed to give the height of the peak of the CSF curve. The chart is based on the theory that abnormal CSFs can be treated as if they are normal CSFs that are shifted to the left (reduced acuity) or down (reduced sensitivity). This method relies on the controversial assumption that notches and other departures from the parabolic shape of the CSF are either nonexistent or at least of no clinical significance (20, 37). Another assumption is that the peak CSF is the best parameter for visual function. The Pelli-Robson chart is read from a standard distance and presents single-size letters, grouped in triplets. Contrast is reduced with each set of triplets in equal logarithmic steps. Contrast sensitivity is measured as the last triplet that the subject read two out of three correctly.

Choosing a Glare Source

Glare sources act to degrade the contrast of the stimulus. The reduction is related to the position and intensity of the glare source as well as the light-scattering properties of the visual system. The optics of the eye tends to focus a point light source, such as oncoming headlights, at the macula. Subjects tested using point sources, however, tend to be distracted by the direct light in their visual periphery and break fixation periodically to stare into the light. Diffuse light sources tend to irradiate the retina in a more homogenous fashion (Figure 5). Since we do not look directly at the sun but rather are exposed to sunlight reflected off surfaces, diffuse lighting studies are a more practical test for daytime lighting (38).

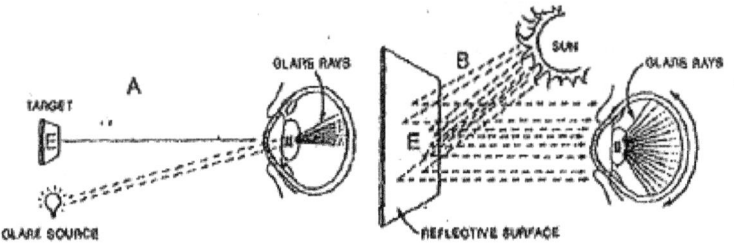

Figure 5: *point(A)* vs. *diffuse (B) glare source (19)*

Current glare testers are usually associated with standard Snellen visual acuity charts or contrast sensitivity charts. The Miller-Nadler glare tester attempts to illuminate the background to the level of "new snow on a sunny day" while testing for contrast sensitivity with Landolt C's. The Brightness Acuity Tester (BAT) provides a uniform glare source by projecting a hemisphere of light under the eye while reading a Snellen chart. The Vistech VCT8000 puts a reduced-size Vistech chart in a viewing system that controls illumination. The EyeCon 5 is a PC program that provides constant luminance in conjunction with Landolt C's or bar gratings in different orientations (19).

In the current literature, stimulus and response parameters as well as the type of glare source vary extensively; there is a lack of consistency in methodology and equipment. Glare testing with Snellen charts showed better prediction of outdoor visual acuity but a comparison between testers gave varying results depending on the type and instead completed our study in the sunny conditions that athletes typically find location of the opacity (39). For this reason, we chose not to use a glare tester and themselves playing. However, in future studies, nighttime stadium lighting may be a more controlled source of glare.

Testing Standardization

Testers who ask subjects to read as many letters as they can will get a different result from one who asks the same subjects to guess at hard-to-see letters. First-time subjects also tend to be conservative in their answers; requiring them to guess improves measuring accuracy (20). We standardized our testing protocol by having a single tester test all subjects for each chart. Each tester asked the subject to continue guessing until the tester recorded two incorrect responses within a triplet.

Statistical Analysis

Prior to testing, preliminary sample size calculations, based on an « of 0.5 and power of 80, indicated that we would need a sample size of 40 subjects in each treatment group to have significant results. Our sample size was much smaller, with a total of 46 participants. As a result, it was unclear whether our results would achieve the normal distribution necessary for parametric analysis. We analyzed the data with both parametric and non-parametric tests: parametric analysis was done with ANOVA and paired-T tests, equivalent non-parametric tests were the Kruskal-Wallis one way ANOVA and the Wilcoxon matched-pairs signed-ranks test. Our results were nearly identical in either case, indicating that our small sample size nevertheless reflected a normal distribution. Thus we present the data here with parametric analysis.

Bias

It was impossible to avoid a certain level of bias associated with the application of the various products. EyeBlack is prevalent enough in culture that participants could guess the intended effects. We could think of no way to blind them to their treatment groups. We also suspect a learning bias based on repeated readings of the chart. Subjects were asked to read both charts with each eye separately and then binocularly. In all, each subject gave six readings, three for each chart. We attempted to control for this possibility by providing charts with different letters. However, we found that binocular scores were consistently higher than monocular scores; whether this is the effect of longer exposure to the chart that resulted in cladfication of hard-to-see letters or the superiority of binocular contrast sensitivity is difficult to assess.

Demographic Variables

Sports commonly associated with EyeBlack use are football and baseball. Current demographics of these groups tend to be all male with equal parts light and dark skinned players. It is notable that EyeBlack is not used in tennis or soccer or many other outdoor sports whose demographics are more skewed to Caucasian races. It is also worth mentioning that EyeBlack was first introduced many years ago as burnt cork placed on the face of mainly white players.

A large percent of our sample was Caucasian and we had no African American participants. Fair-skinned individuals should theoretically benefit more from EyeBlack since the product effectively increases the concentration of pigment that would absorb light on the cheekbone. Dark skin and light have essentially the same number of melanocytes; however, total melanin concentrations in albino skin is nearly zero while the entire body melanin concentration of a black adult is roughly one gram (40). Studies have shown that increased concentration of melanin in intraocular structures help reduce light damage by absorbing more incident light and scavenging for free radicals (38). Dark skin has been epidemiologically shown to be protective to certain UV exposure, most likely because of increased melanin concentration in the eye (38). As mentioned earlier, there are race differences associated with contrast sensitivity and stimuli orientation with Caucasians showing decreased sensitivity to oblique orientations (oblique effect) (25). However, we found no studies to date on athletes, race differences, and the perception of glare.

Our sample also had a slight majority of women, which introduces the question as to whether it would be possible that EyeBlack benefits male players. Brabyn and McGuinness (41) have shown there are gender differences in contrast sensitivity, with women more sensitive to low spatial frequencies and men more sensitive to high spatial frequencies but no significant difference in midrange spatial frequencies. Our study itself was too small to give any reliable information on gender breakdown.

Lastly, the sample was not selected for athletic superiority. Laby et al. (3) and many others have shown that athletes have superior visual skills than nonathletes. Significant differences are found between population norms, major league, minor league, and collegiate ball players in multiple visual parameters, including contrast sensitivity. Most participants in out study indicated some level of sport activity ranging from recreational to college varsity. The event itself from which participants were recruited was an athletic event and probably selected for athletic individuals.

Binocular vs. Monocular Testing

There is questionable validity in combining right and left eye data to double the sample size. It has been shown that left and right eye readings are not independent readings, which stem from the fact that they arise from the same single brain. In terms of CSF testing, binocular CSF has been shown to have higher sensitivity than monocular sensitivity across all spatial frequencies compared. The difference was shown to be -42% higher than the predicted sum of the monocular responses (42, 43). Gilchrist and McIver (43) also showed that decreased luminance in one eye from any ocular condition decreases the binocular CSF such that it is worse than the better eye. In our own testing, binocular testing was clearly superior to monocular. However, because testing procedure involved testing each eye separately and then both eyes together, it is possible that the longer exposure to the charts contributed to the statistically significant improvement using both eyes.

Ocular Variables

Confounders such as pupil size and visual acuity could also affect contrast sensitivity. Increased pupil size would increase the lens surface area exposed to light. Lens opacities would scatter more light. Visual acuity could also affect testing since defocusing affects letter identification. Participants wearing glasses were excluded from the study because of light scattered from the glass lens. Contact lens users were enrolled in the study but may have increased light scatter secondary to corneal edema. Corneal edema has been shown to have a small effect on contrast sensitivity but may increase glare sensitivity up to 300% (19). Kluka et al. (44) also looked at the effects of daily-wear contact lenses on contrast sensitivity in ten professional and ten collegiate-level female tennis players. She found that female tennis players without corrective lenses had significantly higher contrast sensitivity at intermittent and high spatial frequencies than those who wore daily-wear soft contact lenses.

In this study, we were unable to measure pupil size using a pupillometer because of the bright conditions. We also did not have time in the narrow testing period to test visual acuity. We feel that the Pelli-Robson chart minimizes the effect of refractive error by utilizing large, single-size letters read at a distance of one meter. We found that the mean Pelli-Robson values did not vary much between the control groups (Table 6) regardless of eye tested. We can infer from this that all participants, whether they were randomized to EyeBlack grease, Vaseline grease, or No Glare sticker, were able to perform similarly at baseline.

Other Limitations of Study

Sunlight luminance and the position of the sun was not recorded in this study; as a result, it will be difficult to relate our results with future results achieved in similar settings. Variability in ambient testing conditions could affect our results by exposing participants to varying levels of sun brightness. Perceptual brightness changes with the angle of the sun, with cloud conditions, and with time of year. We achieved our testing within a three-hour period on a single day to minimize variability in sun luminance; however, gradual changes could not be assessed.

Certain other factors of the study would be difficult to control. For instance, it is possible that in actual athletic performance, contrast sensitivity would change, whether as a result of increased adrenergic response or unmeasurable neural activity. In a study measuring contrast sensitivity using sinusoidal gratings before and after jogging, Koskela et al. (45, 46) found that in most of the 11 subjects, there was a statistically significant improvement in contrast sensitivity after jogging in all spatial frequencies tested, suggesting that in some individuals physical exercise triggers factors that affect visual function. Additionally, sweating during sports events may affect glare; beads of sweat that gather on the brow or cheekbone may reflect light into the eye. In both cases, changes in contrast sensitivity would be difficult for us to assess.

Conclusion

We have designed a study to test whether EyeBlack grease and No Glare stickers reduce glare and improve contrast sensitivity during sunlight exposure. In a randomized, controlled trial, we measured participants' contrast sensitivity using the Pelli-Robson chart before and after application of one of three treatment groups: EyeBlack grease, No Glare sticker, and Vaseline placebo. We have found EyeBlack to be statistically superior to control and to No Glare sticker in three situations. There is a statistically significant difference between the EyeBlack grease and No Glare sticker in binocular testing. There is also a statistically significant difference between the control and EyeBlack grease in binocular testing and in the combined data of the right and left eyes. These results suggest that EyeBlack grease does in fact have antiglare properties while No Glare sticker and Vaseline do not.

The greatest challenge facing contrast sensitivity measurement and glare testing is the lack of standardization in procedures, both in stimulus parameters and testing style. We decided to forgo an artificial glare source and instead used natural sunlight. However, we did not measure sunlight luminance and therefore cannot determine if variations in lighting affected our results. Future studies would benefit from a controlled glare source. We then chose to measure contrast sensitivity with the Pelli-Robson contrast chart specifically for its familiarity to participants, its high test-retest reliability, and its ease of function. We still believe that this is the most efficient test available for these conditions; however, we believe there would be value in evaluating the levels of contrast that lie between 1.65 and 1.80.

It is unclear whether the statistical significance may have been affected by the sample size or the demographics of the group. Our small sample size proved particularly problematic at looking at subgroups within the study. We did not have enough participants such that we could evaluate performances between men and women or light and dark skin color. Further testing with larger sample sizes and controlled glare conditions are needed to determine if glare reduction occurs in a manner that would reduce sun glare in actual athletic conditions.

APPENDIX 1

EYEBLACK SURVEY QUESTIONNAIRE

Age   Gender   M   F
Ethnicity
Do you wear contacts   Y   N
Have you ever been treated for an ocular   Y   N
condition?
If yes, please describe:

Do you participate in sports (that require   Y   N
hand-eye coordination)?

At what level? (circle an that apply)
    High School   JV / V
    College: JV /V / Intramural
    Recreational

Which sports do you play?

Have you ever used EyeBlack in the past?   Y   N

APPENDIX 2

PELLI-ROBSON CONTRAST CHART DATA COLLECTION SHEET

PELLI-ROBSON CONTRAST SENSITIVITY TEST

0.00 H S Z D S N 0.15	0.00 H S Z D S N 0.15	0.00 H S Z D S N 0.15
0.30 O K R Z V R 0.45	0.30 O K R Z V R 0.45	0.30 O K R Z V R 0.45
0.60 N D O O S K 0.75	0.60 N D O O S K 0.75	0.60 N D O O S K 0.75
0.90 O Z K V H Z 1.05	0.90 O Z K V H Z 1.05	0.90 O Z K V H Z 1.05
1.20 N H O N R D 1.35	1.20 N H O N R D 1.35	1.20 N H O N R D 1.35
1.50 V R O O V H 1.65	1.50 V R O O V H 1.65	1.50 V R O O V H 1.65
1.80 O D S N D O 1.95	1.80 O D S N D O 1.95	1.80 O D S N D O 1.95
2.10 K V Z O H R 2.25	2.10 K V Z O H R 2.25	2.10 K V Z O H R 2.25

Right Eye   Binocular   Left Eye

Log Contrast Sensitivity: _____ Log Contrast Sensitivity: _____ Log Contrast Sensitivity: _____
Acuity: _____ Acuity: _____ Acuity: _____
Correction: _____ Correction: _____
Pupil Diameter: _____ mm Pupil Diameter: _____ mm

Name: _____ Comments: _____
Age, Sex: _____
Diagnosis: _____
Medications: _____
Date: _____
Examiner: _____

PELLI-ROBSON CONTRAST SENSITIVITY TEST

0.00 V R S K D R 0.15	0.00 V R S K D R 0.15	0.00 V R S K D R 0.15
0.30 N H O S O K 0.45	0.30 N H O S O K 0.45	0.30 N H O S O K 0.45
0.60 S O N O Z V 0.75	0.60 S O N O Z V 0.75	0.60 S O N O Z V 0.75
0.90 O N H Z O K 1.05	0.90 O N H Z O K 1.05	0.90 O N H Z O K 1.05
1.20 N O D V H R 1.35	1.20 N O D V H R 1.35	1.20 N O D V H R 1.35
1.50 O D N Z S V 1.65	1.50 O D N Z S V 1.65	1.50 O D N Z S V 1.65
1.80 K O H O D K 1.95	1.80 K O H O D K 1.95	1.80 K O H O D K 1.95
2.10 R S Z H V R 2.25	2.10 R S Z H V R 2.25	2.10 R S Z H V R 2.25

Right Eye   Binocular   Left Eye

Log Contrast Sensitivity: _____ Log Contrast Sensitivity: _____ Log Contrast Sensitivity: _____
Acuity: _____ Acuity: _____ Acuity: _____
Correction: _____ Correction: _____
Pupil Diameter: _____ mm Pupil Diameter: _____ mm

Name: _____ Comments: _____
Age, Sex: _____
Diagnosis: _____
Medications: _____
Date: _____
Examiner: _____

References

1. Christenson GN and Winkelstein AM. 1988. Visual skills of athletes versus nonathletes: Development of a sports vision testing battery. *Journal of the American Optometry Association.* 59(9): 666–675.
2. Hitzeman SA and Beckerman SA. 1993. What the literature says about sports vision. *Optometry Clinics.* 3(1): 145–169.
3. Laby DM, Rosenbaum AL, Kirschen DG, Davidson JL, Rosenbaum D, Strasser C, and Mellman MF. 1996. The visual function of professional baseball players. *American Journal of Ophthalmology.* 122: 476–485.
4. Stine CD, Arterburn MR, and Stern NS. 1982. Vision and sports: A review of the literature. *Journal of the American Optometry Association.* 53(8): 627–633.
5. Solomon H, Zinn WJ, and Vacroux A. 1988. Dynamic stereoacuity: A test for hitting a baseball? *Journal of the American Optometry Association.* 59: 522–526.
6. Kioumourtzoglou E, Kourtessis T, Michalopoulou M, and Derri V. 1998. Differences in several perceptual abilities between experts and novices in basketball, volleyball, and water-polo. *Perceptual and Motor Skills.* 86: 899–912.
7. Melcher MH and Lund DR. 1992. Sports vision and the high school student athlete. *Journal of the American Optometric Association.* 63(7): 466–474.
8. Hoffman LG, Polan G, and Powell J. 1984. The relationship of contrast sensitivity functions to sports vision. *Journal of the American Optometric Association.* 55(10): 747–752.
9. Kluka DA, Love PL, Kuhlman J, Hammach G, and Wesson M. 1996. The effect of a visual skills training program on selected collegiate volleyball athletes. *International Journal of Sports Vision.* 3(1): 23–34.
10. Campbell FW, Rothwell SE, and Perry MJ. 1987. Bad light stops play. *Ophthalmic and Physiological Optics.* 7(2): 165–167.
11. Perry MJ, Campbell FW, and Rothwell SE. 1987. A physiological phenomenon and its implications for lighting design. *Lighting Research and Technology.* 19: 1–5.
12. Rothwell SE and Campbell FW. 1987. The physiological basis for the sensation of gloom: Quantitative and qualitative aspects. *Ophthalmic and Physiological Optics.* 7(2): 161–163.
13. Rubin GS. 1992. Clinical Glare Testing. In *Current Practice in Ophthalmology.* AP Schachat, ed. St. Louis: Mosby Year Book. 153–163.

14. Miller D and Sanghvi S. 1990. Contrast Sensitivity and Glare Testing in Corneal Disease. In *Glare and Contrast Sensitivity for Clinicians*. MP Nadler, D Miller, and DJ Nadler, eds. New York: Springer-Verlag. 45–52.
15. Lerman S. 1987. Light-Induced Changes in Ocular Tissues. In *Clinical Light Damage to the Eye*. D Miller, ed. New York: Springer-Verlag. 183–215.
16. Chatman WN and MacEwen CJ. 1995. Light and lighting. In *Sports Vision*. DFC Loran and CJ MacEwen, eds. Oxford: Butterworth-Heinemann. 88–112.
17. Miller D. 1991. Light Damage to the Eye. In *Textbook of Ophthalmology*. SM Podos and M Yanoff, eds. New York: Gowers Medical Publishing. 2.4.1–2.4.6
18. Vinger, PF. 1994. The eye and sports medicine. In *Duane's Clinical Ophthalmology, Volume 5*. W Tasman and EA Jaeger, eds. 1–103.
19. Prager TC. 1990. Essential Factors in Testing for Glare. In *Glare and Contrast Sensitivity for Clinicians*. MP Nadler, D Miller, and DJ Nadler, eds. New York: Springer-Verlag. 33–44.
20. Wolfe JM. 1990. An Introduction to Contrast Sensitivity Testing. In *Glare and Contrast Sensitivity for Clinicians*. MP Nadler, D Miller, and DJ Nadler, editors. New York: Springer-Verlag. 5–23.
21. Bodis-Wollner 1. 1972. Visual acuity and contrast sensitivity in patients with cerebral lesions. *Science*. 178(62): 769–771.
22. Bodis-Wollner I and Diamond SP. 1976. The measurement of spatial contrast sensitivity in cases of blurred vision associated with cerebral lesions. *Brain*. 99(4): 695–710.
23. Schwartz SH. 1994. *Visual Perception: A Clinical Orientation*. Norwalk, CT: Appleton and Lange. 384 pp.
24. Storch RL and Bodis-Wollner 1. 1990. Overview of Contrast Sensitivity and Neuroophthalmic Disease. In *Glare and Contrast Sensitivity for Clinicians*. MP Nadler, D Miller, and DJ Nadler, eds. New York: Springer-Verlag. 85–112.
25. Appelle S. 1972. Perception and discrimination as a function of stimulus orientation: The "oblique effect" in man and animals. *Psychology Bulletin*. 78(4): 266–278.
26. Miller D. 1974. The effect of sunglasses on the visual mechanism. *Survey of Ophthalmology*. 19(1): 38–44.
27. Miller D and Nadler MP. 1990. Light Scattering: Its Relationship to Glare and Contrast in Patients and Normal Subjects. In *Glare and Contrast Sensitivity for 44 Clinicians*. MP Nadler, D Miller, and DJ Nadler, eds. New York: Springer-Verlag. 24–32.

28. Sliney DR. 2001. Photoprotection of the eye—UV radiation and sunglasses. *Journal of Photochemistry and Photobiology B: Biology.* 64(2001): 166–175.
29. Vinger PF. 1996. Introduction in *Sports Ophthalmology.* BM Zagelbaum, ed. Cambridge: Blackwell Science. 1–22.
30. Stephens GL and Davis JK. 2000. Spectacle Lenses. In *Duane's Ophthalmology Volume 1.* W Tasman, ed. Philadelphia: Lippincott, Williams, and Wilkins. Chapter 51. 1–64.
31. Salisbury JA. 1987. The Eye and Shooting. In *Sports Ophthalmology.* LD Pizzarello and BG Haik, eds. Springfield, IL: Charles C. Thomas. 139–148.
32. Marmor MF. 2001. Double Fault! Ocular Hazards of a Tennis Sunglass. *Archives of Ophthalmology.* 119: 1064–1066.
33. Bryce R. 2001. Out of left field. www.allstinchronicle.com/issues/voI118/isslle44/xtra.leftfield.html.
34. Glareblox stick on strips product website. www.glareblox.com/historyl/html.
35. Elliot DB, Sanderson K, and Conkey A. 1990. The reliability of the Pelli-Robson contrast sensitivity chart. *Ophthalmic and Physiological Optics.* 10(1): 21–24.
36. Lempert P. 1990. Standards for Contrast Acuity/Sensitivity and Glare Testing. In *Glare and Contrast Sensitivity for Clinicians.* MP Nadler, D Miller, and DJ Nadler, eds. New York: Springer-Verlag. 113–119.
37. Pelli DO, Robson JG, and Wilkins AJ. 1988. The design of a new letter chart for measuring contrast sensitivity. *Clinical Vision Science.* 2(3): 187–199.
38. Weiter J. 1987. Phototoxic Changes in the Retina. In *Clinical Light Damage to the Eye.* D Miller, ed. New York: Springer-Verlag. 79–126.
39. Nadler DJ. 1990. Glare and Contrast Sensitivity in Cataracts and Pseduophakia. In *Glare and Contrast Sensitivity for Clinicians.* MP Nadler, D Miller, and DJ Nadler, editors. New York: Springer-Verlag. 53–65.
40. Miller D and Stegmann R. 1987. Approaches to Protection Against Light-Induced Changes in the Eye. In *Clinical Light Damage to the Eye.* D Miller, ed. New York: Springer-Verlag. 165–179. 45
41. Brabyn LB and McGuinness D. 1979. Gender differences in response to spatial frequency and stimulus orientation. *Perceptive Psychophysiology.* 26(4): 319.
42. Leege GE. 1984. Binocular contrast summation: Detection and discrimination. *Vision Research.* 24(4): 373.

43. Gilchrist J and McIver C. 1985. Fechner's paradox in binocular contrast sensitivity. *Vision Research*. 25(4): 609.
44. Kluka DA and Love PA. 1993. The effects of daily-wear contact lenses on contrast sensitivity in selected professional and collegiate female tennis players. *Journal of the American Optometry Association*. 64(3): 182–186.
45. Koskela PD. 1988. Jogging and contrast sensitivity. *Acta Ophthalmologica (Copenhagen)*. 66(6): 725–727.
46. Koskela PD, Airaksinen PJ, and Tuulonen A. 1990. The effect of jogging on visual field indices. *Acta Ophthalmologica (Copenhagen)*. 68(1): 91–93.
47. Beckerman SA and Hitzeman S. 2001. The ocular and visual characteristics of an athletic population. *Optometry*. 72(8):498–509; Berman AM. 1993. Clinical evaluation of the athlete. *Optometry Clinics*. 3(1):1–26.
48. Classe JG. 1993. Prescribing for noncontact sports. *Optometry Clinics*. 3(1):111–128.
49. Davis JK. 1987. Lenses for Sports Vision. In *Sports Ophthalmology*. LD Pizzarello and BG Haik, eds. Springfield, IL: Charles C. Thomas. 9–43.
50. Home R. 1978. Binocular summation: A study of contrast sensitivity, visual acuity, and recognition. *Vision Research*. 18: 579–585.
51. Leat SJ and Woo GC. 1997. The validity of current clinical tests of contrast sensitivity and their ability to predict reading speed in low vision. *Eye*. 11: 893–899.

www.ingramcontent.com/pod-product-compliance
Lightning Source LLC
Chambersburg PA
CBHW031215090426
42736CB00009B/922